Early Praise for *Essential 555 IC*

I'd recommend it to someone beginning in digital electronics, for sure. Having actual projects and troubleshooting is great. Having further tinkering is a big bonus.

➤ **Caleb Kraft**
   Senior Editor, Make Community

I like that the book does not make too many assumptions about the reader. I also like how the author provides a framework for the reader to expand upon each project to further their understanding.

➤ **Drew Fustini**
   Engineer, OSH Park

Essential 555 IC

Design, Configure, and Create Clever Circuits

Cabe Force Satalic Atwell

The Pragmatic Bookshelf

Raleigh, North Carolina

For our complete catalog of hands-on, practical, and Pragmatic content for software developers, please visit *https://pragprog.com*.

The team that produced this book includes:

CEO: Dave Rankin
COO: Janet Furlow
Managing Editor: Tammy Coron
Development Editor: Patrick Di Justo
Copy Editor: L. Sakhi MacMillan
Layout: Gilson Graphics
Founders: Andy Hunt and Dave Thomas

For sales, volume licensing, and support, please contact *support@pragprog.com*.

For international rights, please contact *rights@pragprog.com*.

ISBN-13: 978-1-68050-783-6
Book version: P1.0—April 2021

Contents

Acknowledgments

I'd like to thank everyone who helped me to create this book, including Tammy Coron, Janet Furlow, Andy Hunt, Patrick Di Justo, and Dave Rankin at Pragmatic. Special thanks to tech editor Caleb Kraft.

I couldn't have made this book without Lydia, my wife, my love, my friend.

I dedicate this book to my son Remy...someone embarking down the road of building and learning.

Preface

The venerable 555 Timer Integrated Circuit (IC) is a little chip of immense power. A half-century-old, this IC shines like a star in a sea of electronic components. Equipped with the 555 Timer, in combination with this tomb of knowledge, anything is possible, and nothing stands in the way.

Yes, the 555 Timer is like a little magical item. It can flash a light, make a noise, store data, play music and more. Its capabilities are limited only by imagination.

Rare is the knowledge to wield the 555 Timer. But, it is not exactly a rare magical item. Billions are made every year, and billions are used all around the world.

Whether you are a hobbyist, beginner or seasoned engineer… learning how to use a 555 Timer is the real magic, and it can go a long way in your design life.

Who should read this book?

From beginner to novice to expert, everyone can benefit from knowing how to use 555 Timers to solve design problems. In just a few minutes, you'll build your first project and see what the 555 can do. By the time you finish the book, you'll be able to build interesting and useful 555 Timer circuits.

If you're a novice, already familiar with circuit building, this book gives you new tools, tricks and designs to lead you further down the road of your journey.

If you are a bit of an expert already, this book undoubtedly adds something new to your catalog of useful techniques.

Using the chip is fun, informative and insightful. Knowing the foundational use of the timer IC goes a long way in developing products, projects or fun toys.

What's in this book?

This book is a collection of projects, through which every aspect of the 555 Timer IC is touched upon. The gradual cascading of complexity in each project further expands on the reader's knowledge of and use of the 555 Timer.

How to read this book.

Feel free to skip around. Each chapter is a self-contained project. Although building the projects in order increases complexity as it goes on, in the way of easing into different features of the 555 Timer, but it isn't mandatory to do so.

I do think that going project by project comfortably eases the novice into building circuits, and refresh those who know at a reasonable pace. Each project is useful and benefits the reader's understanding. The hands-on nature of the book translates into instant retention of the techniques.

Turning on a single LED might seem like a skippable project. But, if you can turn on a LED, you can have it make a sound – have it turn on another circuit – have it trigger another project. Each project is much more than it seems.

The basic concepts cascade through the projects like this:

- Turning on a single LED for a specific amount of time.
- Flashing LEDs, opposing each other.
- Creating a chasing effect on a string of LEDs.
- Using timing to create a traffic light system.
- Making a controllable dimmer circuit.
- Creating a silent alarm system or storing a single bit of memory.
- Creating a sound synthesizer and beyond.

Conventions of the book.

Another useful feature of the book for the beginner and novice is understanding how to read a circuit diagram and a breadboard layout. The circuit diagram is the map, the exact instructions on how to build the circuit.

Throughout the book, the reader builds those circuits on a breadboard. This component lets the user sort of build the circuits like they would LEGO bricks. They push the electronic component into the breadboard to start forming the circuit.

Using a breadboard is important for the beginner to understand its functionality. This is how they can build an actual circuit without a soldering iron or industrial oven. It's also a speedy way of building and testing circuits too.

Online resources.

At the official website for this book, https://forum.devtalk.com/t/essential-555-ic-pragprog/1168, you'll find the following:

- A discussion forum where you can communicate directly with the author and other readers.
- An errata page, listing any mistakes in the current edition (let's hope that list will be empty!)

Let's begin!

CHAPTER 1

Getting Started

The 555 timer integrated circuit (IC) is an early multipurpose electronic device, used by manufacturers and hobbyists for decades. It's the most popular IC ever made—billions of 555 timer chips are in existence, and over 900 million new chips are produced every year.

Three Main Uses for the 555

1. As a monostable one-shot *pulse generator*. In this mode, the 555 can be a timer, a touch switch, a frequency divider, and more.

2. As a bistable *flip-flop*. These devices are the fundamental building blocks of digital computation—they have the ability to hold, set, reset, or toggle one *bit* of information, depending on how they're wired up.

3. As an astable *oscillator*. This configuration causes the 555 to repeatedly output an electronic pulse. This can be used to flash a light, create an audio tone, or even convert an analog signal to a digital one.

This book will give examples of all three modes.

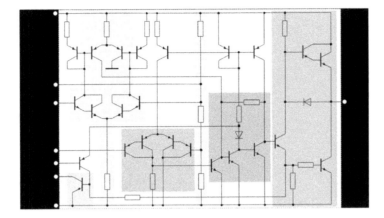

This Is the Integrated Circuit Inside a 555 Timer! It's Packed!

The 555 has a few dozen electronic components squished into a tiny package. In fact, that's why little chips like these are called *integrated* circuits: all the pieces needed to make this circuit are "integrated" into a convenient package, so you don't have to build the circuit over and over again.

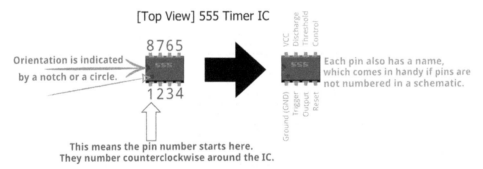

The preceding line drawing shows how all the individual 555 pins are labeled. Knowing what they all do isn't important for now, but you will need to know what the pins are called so you can hook them up correctly.

In this book you will find all the projects started as a line art schematic. The traditional way a circuit design would be is shown in the line drawing on page 3.

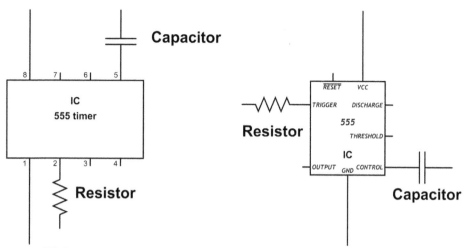

This is an electrical schematic showing two different ways 555 timers can be depicted.

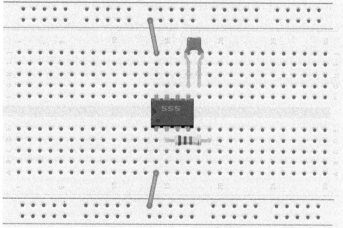

This is the same build as the schematic.

The schematic will be followed by a breadboard view illustration. It is exactly the same design as in the schematic, but it's assembled on the breadboard in a way that best suits the physical space.

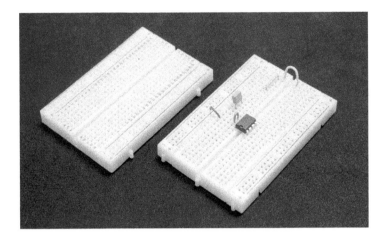

This is a real-world image of a solderless breadboard with the same 555 timer circuit installed on it. The support components from the circuit are placed on it. The real-world build and the breadboard illustration will be fairly close in design, making it useful for building and troubleshooting.

Most solderless breadboards have a center groove that acts as a spacing between the top and bottom halves of the board. One reason is to support an integrated circuit. It's often referred to as the ravine or channel. All similarly built IC chips can bridge this gap as they are plugged into the breadboard. This is the easiest way to use the 555 timer.

555 timers don't care too much about the voltage you give them. Usually, as long as it's somewhere between 4.5 and 15 volts, the 555s work happily. You can power your projects with three AA batteries, a couple of 3 volt CR2032 coin-cell batteries, a 9 volt battery, 4 LiPo power packs, or anything in between.

This is the top and bottom of a 555 timer IC.

Take close a look at the "legs" of the 555. Two rows of four pins connect to the long edges of the chip. These pins are arranged this way on purpose, in what is called a *Dual In-Line Package*, or DIP for short. In this case, the eight pins of the 555 timer are in a DIP8 configuration. Manufacturers have agreed on a standard spacing between all DIP chips' pins, and they fit perfectly into breadboards.

Anode and Cathode

In this book you're going to hear quite a bit about *anodes* and *cathodes*. In the field of electronics, an anode is a positively charged electrode and a cathode is a negatively charged electrode. For some of the components we deal with in this book, such as speakers, resistors, switches, and potentiometers, it doesn't really matter which end of the component is positive and which end is negative. But LEDs will only work correctly if they're hooked up the right way. To make it simple, all LEDs come from the manufacturer with one long leg and one short leg. The long leg of the LED is the anode, and the short leg is the cathode. You can remember it easily by thinking that the long leg has had some length added to it (making it a +), and the short leg has had length subtracted from it (making it a –).

Now that you know what a 555 timer IC is and how we'll build projects in this book, we'll jump right into the first project.

CHAPTER 2

One-Shot

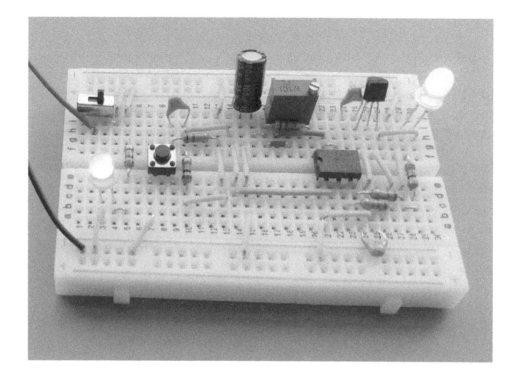

Pressing a button to turn on a light—does it have to be this complex?

One-Shot. One time and done. The name says it all.

In this project, you'll press a button and the circuit will trigger once. Exactly once. You can press the button as many times as you want after that, and it won't make a difference. The circuit stays triggered. This circuit could be used to trigger any sort of action: a siren, beeper, or noise circuit. For this particular project, you'll start with just a single LED.

Parts

555 Timer IC	Newark part number 58K8943	Jameco part number 27423
An LED, any color. I used yellow and green.	Newark part number 97K4048 and 97K4041	Jameco part number 334108 and 693901
NPN Transistor, 2N3904	Newark part number 83C3116	Jameco part number 178597
220 Ω Resistor	Newark part number 38K0351	Jameco part number 690700
4.7 kΩ Resistor	Newark part number 38K0304	Jameco part number 691024
10 kΩ Resistor	Newark part number 38K0328	Jameco part number 691104
2x 15 kΩ Resistor	Newark part number 58K5016	Jameco part number 691147
150 kΩ Resistor	Newark part number 58K5017	Jameco part number 691382
100 kΩ Potentiometer	Digikey part number CF14JT100KCT-ND	Jameco part number 853599
2x 0.01 µF 10 nF Ceramic Capacitor	Newark part number 46P6665	Jameco part number 15229
22 µF Electrolytic Capacitor	Newark part number 69K7919	Jameco part number 1946295
9 V Battery Holder/Strap	Newark part number 31C0662 or 59K0356	Jameco part number 101470
9 V Battery	Newark part number 81F157	Jameco part number 27423
Momentary Contact Tactile Switch	Newark part number 60M5365	Jameco part number 153252
Slider Switch SPDT	Newark part number 10X9279	Jameco part number 2192384
Half-size Breadboard	Newark part number 99W1759	Jameco part number 2157693

A side note about alternative parts and the bill of materials, the BOM:

Although I do provide a list of parts and part numbers, there's a chance the supplier is sold out or you don't want to buy just one 1 kOhm resistor for the project at hand. It's OK, to a limit. Be forewarned, you can sway on the values by a handful of percent, and the project may work. Jaunt too far in the value fields, and the project will stop working properly or at all in some cases.

Adding a big resistor, for example, is like pulling a wire out. Air has a resistance, a super-high resistance, like that one you want to add. Adding a bigger capacitor is like adding a super-large gas tank to your car, in some cases. You'll drive forever, just like the LED pulling power from that big capacitor you want to add.

Although not that common, some breadboards will differ from the ones I suggest. So connections will not be accurate. Different slider switches might operate differently than my suggestions. Be aware.

If, for some reason, you select different LEDs than in my list, make sure they have the same forward voltage and current draw.

The Official Schematic

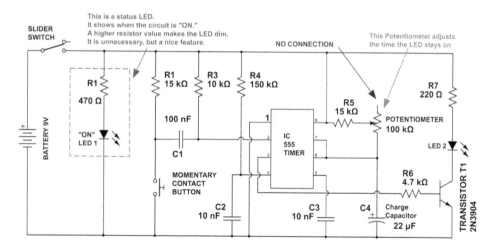

This schematic adds an optional LED that shows when the circuit is on. It is shown within the red square. It's completely optional.

You're setting the 555 timer IC in a state called *monostable multivibrator*, or just monostable for short. This is sometimes called a *one-shot* circuit.

How this works in brief: When the battery is connected to this circuit, pin 2 is set high. When you press the button (which is just a momentary contact switch), pin 2 is set low and pin 3 is set high. Pin 3 turns on the transistor, and in turn the LED.

Note the potentiometer and the capacitor in the circuit. Together they determine how long the LED stays on. When the one-shot is activated, the capacitor begins charging. The resistance of the potentiometer determines how fast the capacitor fills. Once the charge in the capacitor reaches a certain voltage, pin 3 turns off.

This resistor and capacitor combination is often called an RC Circuit. The product of the value of R and the value of C is the time constant: the time that pin 3 stays energized. This is defined by an equation:

(Time = 1.1 * R * C)

The formula reads:

Time in seconds = 1.1 x Potentiometer in ohms x Charge Capacitor in Farads

If you set the potentiometer to the maximum of 100 kOhm, the formula would work out as follows:

Time = 1.1 x 100,000 (ohms) x 0.000022 (Farads)

Time = 2.42 seconds

If you want the LED to be on longer than this circuit is designed for, all you need to do is to increase the value of the capacitor.

Breadboard View

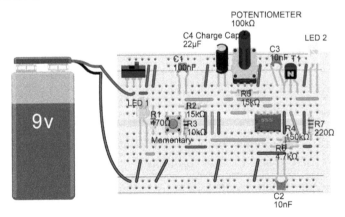

Up close and personal with a 555 timer IC.

Make sure to place the momentary contact button somewhere on the breadboard where you can easily access it.

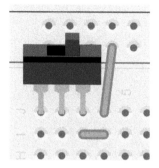

Place the power switch on the breadboard and connect one end to the power bus.

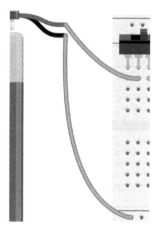

Connect the 9 V battery leads to the switch and GND connections on the breadboard.

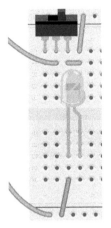

Place the status LED on the board and connect the cathode to GND.

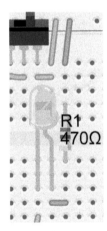

Place the current-limiting resistor and connect it to the power supply and the LED anode.

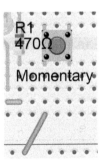

Place the momentary switch on the board and connect one side to GND.

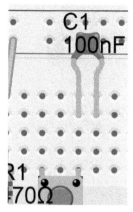

Place the 100 nF capacitor on the breadboard.

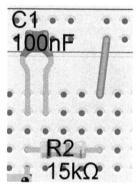

Place the 15 kOhm resistor on the board and connect it to the power.

Place the 10 kOhm resistor on the board and connect it to power.

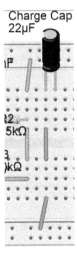

Place the 22 µF charge capacitor on the breadboard and connect the cathode to GND.

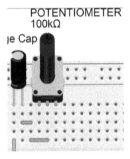

Place the potentiometer on the breadboard and connect it to the 22 µF charge capacitor.

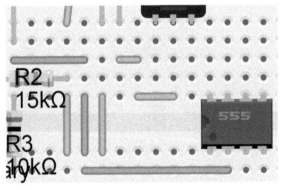

Place the 555 timer on the breadboard, connect pin 2 on the IC to the 100 nF capacitor and 10 kOhm resistor.

Place a 15 kOhm resistor connecting pin 8 on the 555 timer to the middle pin of the potentiometer. Note how that connection is routed from the left of the 15 kOhm resistor to the middle pin with a jumper wire two rows below it on the breadboard.

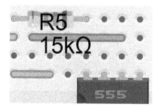

Connect pin 6 on the 555 timer IC to the first potentiometer pin, or leftmost pin.

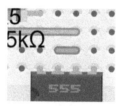

Connect pins 6 and 7 on the 555 timer together.

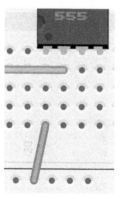

Connect pin 1 on the 555 timer to GND.

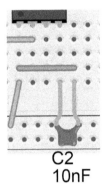

C2
10nF

Connect pin 4 on the 555 timer to 10 nF capacitor and connect the capacitor to GND.

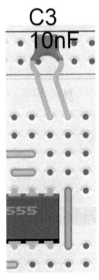

C3
10nF

Connect pin 5 on the 555 timer to a 10 nF capacitor (C3) that will connect to GND net through a wire.

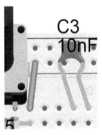

C3
10nF

Connect pin 8 on the 555 timer to power on the breadboard.

Place the 150 kOhm resistor to pin 4 of the 555 timer IC. This will later be connected to power on the breadboard.

Place the transistor and the (R6) 4.7 kOhm resistor. The 4.7 kOhm resistor will connect to pin 3 of the 555 timer and the base of the transistor through the blue wire in this picture.

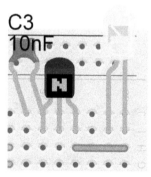

Place LED and connect the LED's cathode (negative lead) to transistor's collector.

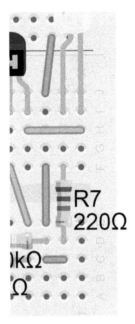

Place a current-limiting resistor for LED on the breadboard and connect to power.

You may have noticed that the potentiometer doesn't have a jumper wire bridging two of the pins together. It's perfectly fine to wire it in this fashion.

Since this project turns on a light for a defined time and turns it off, it acts like a stopwatch. You could use this circuit as a stopwatch, or digital hourglass, and when the light turns off, time is over. Anything where exact time is a factor could apply here. It could be a one-shot style night-light, but our LED might be a little too bright for nighttime eyes.

Button not pressed.

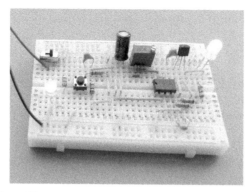

Button pressed, One-Shot activated!

Takeaways

- What did you learn from using a 555 timer in monostable mode? How else would you make a motion detector, or a night-light that stays on for a little bit

- Try adjusting time delays in a one-shot circuit. It's like a simple form of memory. Keep a light on or sound playing long after you pressed the button.

- How would you need to change the circuit so that the LED is on until you press the button to turn it off?

Things to Try

- This project makes a pretty good temporary night light—that is, if you remove the status LED. You can adjust the time the triggered light stays on as you see fit. But if you want it to stay on longer, you can change the size of the charge capacitor. Get one that is twice the size, for example.

- This circuit could technically turn on another circuit if the other was wired up between positive voltage of the battery and the collector of the transistor. So this position acts like the battery of the other circuit, functionally. But a single 9 V battery can't power too many circuits at the same time. I'd remove the status LED in this case.

Troubleshooting

Every chapter will have a troubleshooting section with particular points to pay attention to with the circuit build. Here's the first.

- In this one-shot circuit, pay special attention to all the short wires. There are a lot of them in this design. It's sometimes hard to see where they are being placed. This will most likely be where errors occur.

- Also, another big issue with LED projects is the polarity of the LED's connections. Make sure they are correctly oriented to the voltage source and ground accordingly.

These points are very specific to the one-shot circuit, but there's far more to know about troubleshooting in general. So this next section will go over everything about troubleshooting for this book's, and most other, breadboard projects.

In a perfect world, your circuit is now done and lighting up for a variable amount of time. More often than not, circuit builds don't go that easy. I often liken putting together electronics to putting together LEGO bricks. If you follow the steps, it'll work. I might need to retool that statement. Wires are so small, the places to put them so small that the human eye doesn't perceive them—nothing like LEGO bricks.

You may have to troubleshoot a project or two in this book. That's OK. It's essential on your way to learning 555 timer circuit design.

Every single wire and component placed in these designs is absolutely essential. It will not work without them all. So if it isn't working now, then the issue is simple—they're not all connected.

If you're building these on a breadboard, as I recommend in each project, the most common problem is you've pushed a wire into the wrong spot. Often it's just one space off.

If the battery you're using is new, then proceed down the many levels of the troubleshooting inferno that follows.

- First step, go over every single wire placement. If you built it exactly like my designs, then you have a reference. Go over each spot, compare. Before too long, I bet you'll find one or more wires misplaced.

If you didn't follow my exact builds, just make sure each wire and component at each junction are connected.

- Still not working? OK, the next step is to make sure each wire is pushed into the breadboard. Sometimes it might look like it's connected but isn't.

- Oh no, still not working? Here's where the troubleshooting takes a technical turn. At this point, you're going to want to make sure none of the wires have broken and feel tight inside the breadboard. This can be tedious to do, but a simple touch or wiggle of wire is all you may have to do. Tweezers help here.

- Here's where you may get frustrated. It isn't your fault. This is where we can blame other things! The first one to worry about is the power switch. In all these projects, I used a slider switch. Bypass it, and place the positive power from the battery to the breadboard's positive power bus. At the top of each board, the long horizontal line marked with red is the positive power bus/rail.

When you do this, see if the circuit works. If so, it's the switch. Throw that one away, replace.

- No success yet? Here's where things get serious... It may be time to replace the 555 timer IC. Every so often, from time to time, on occasion... the chip doesn't work. Perhaps it got "fried" from a mis-wiring before you checked the wires themselves. That chip gets fried, it's done. It tried it's best, but it shall never work again! Even after a rewire, it will not come back.

Perhaps some electrostatic discharge zapped its pins somewhere between its manufacture and being in your build today. It happens. Replace it. They're literally a dollar a dozen. You may replace it and it still doesn't work. Try a new one. After the fourth new one, I think it's safe to say it isn't the chip. They probably are all fine. Now...the dreaded last step.

- At the final level of the troubleshooting inferno, treachery! It might be the breadboard itself. I've occasionally found that some breadboards are defective for some unknown reason. To be honest, I've only had this issue with really old breadboards. But instead of trying to deduce the breadboard's issue, replace it with a new one.

Often the act of rebuilding the circuit is a constructive, helpful act. The rebuild might reveal something you missed in an earlier troubleshooting step.

If it still doesn't work, rinse and repeat all of the above. Fingers crossed—I hope you never get to this point.

CHAPTER 3

555 Timer LED Flasher

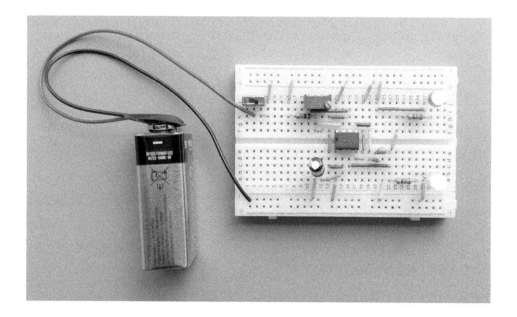

Alternating flashing lights that looks like emergency lights are shown in this image. In this case they are flashing so fast, they both are captured during the camera shutter.

You can configure this project to flash one or two lights, either on-off, or alternating back and forth. This is like an emergency warning blinker, or the lights on top of a police car, and has near-universal recognition. Put this circuit on your bike, on your door, your backpack, you name it.

Parts

555 Timer IC	Newark part number 58K8943	Jameco part number 27423
2x LEDs of your choice.	Newark part number 52K5254 & 04R6674	Jameco part number 206519 & 2234071
2x 470 Ω Resistors	Newark part number 58K5055	Jameco part number 690785
10 kΩ Resistor	Newark part number 38K0328	Jameco part number 691104
100 kΩ Potentiometer	Digikey part number CF14JT100KCT-ND	Jameco part number 853599
100 μF Electrolytic Capacitor	Newark part number 96K9198	Jameco part number 2230539
9 V Battery Holder/Strap	Newark part number 31C0662 or 59K0356	Jameco part number 101470
9 V Battery	Newark part number 81F157	Jameco part number 27423
Slider Switch SPDT	Newark part number 10X9279	Jameco part number 2192384
Half-size Breadboard	Newark part number 99W1759	Jameco part number 2157693

The Official Schematic

The schematic for the LED Flasher is shown on page 25. The 555 timer provides either a positive voltage or a ground for the LEDs, turning them on and off alternately.

This is a 555 timer in the *astable multivibrator* configuration. In this setup, the voltage at the capacitors charges and discharges, oscillating between 66% and 33% of the input voltage. This makes the output voltage at pin 3 vary between 0% and 100% of the input voltage, thereby turning the LED on and off. The cool thing is that the time for charging and discharging doesn't depend on the amount of the input voltage. This circuit will blink at the same rate at 3 volts as it will at 12 volts.

If the voltage doesn't control the charge and discharge time of the circuit, what does? The answer is the value of the resistors and capacitors. The

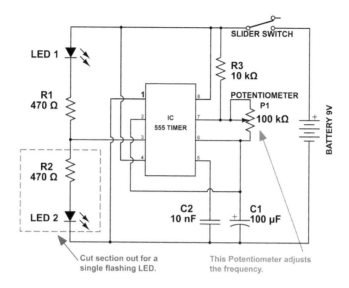

Cut section out for a
single flashing LED.

This Potentiometer adjusts
the frequency.

555 in astable mode has a simple equation to describe the blink rate, or *frequency*:

Charge time = 0.693 * (R1+R2)* C1

Discharge time = 0.693 * R2 * C1

where R1 is the value in ohms of the first resistor, R2 is the value in ohms of the second resistor, and C1 is the value of the capacitor in Farads. The time is measured in seconds.

In this circuit, R1 has a value of 10,000 ohms, and R2 is a potentiometer that can vary from 0 to 100,000 ohms. This means that the two extremes of the circuit can vary from one blink every seven seconds:

Charge time = 0.693 * (10,000 + 100,000) * 0.0001 = 7.6 seconds

Discharge time = 0.693 *100,000 * 0.0001 = 6.9 seconds

to many blinks per second:

Charge time = 0.693 * (10,000 + 1) * 0.0001 = 760 milliseconds

Discharge time = 0.693 * 1 * 0.0001 = 6.9 milliseconds

That makes a very fast blink!

Breadboard View

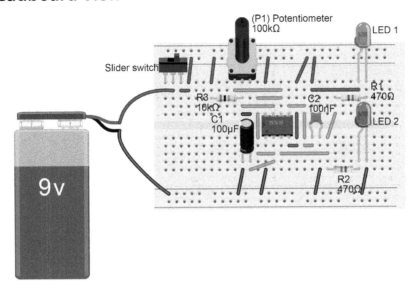

This project only needs a half-size breadboard at best.

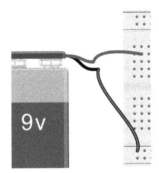

Connect the negative battery lead to the GND bus on the breadboard.

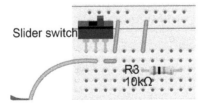

Place the switch onto the breadboard, with one end connected to the positive battery terminal.

Connect the other end of the power switch to the power supply bus.

Place a 10 kOhm resistor on the breadboard and connect one lead of the resistor to the power supply bus.

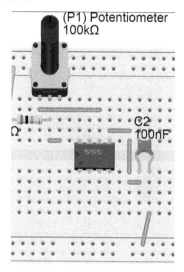

Place the potentiometer on the breadboard and connect the wiper pin to the lead of the 10 kOhm resistor.

Place the 555 timer onto the breadboard.

Connect the bottom lead of the potentiometer to pin 6 on the 555 timer.

Connect pin 7 on the 555 timer to the open lead on the 10 kOhm resistor.

Place the 100 μF electrolytic capacitor on the breadboard and connect one lead to GND.

Connect pin 2 on the 555 timer to the 100 uF capacitor.

Connect pin 8 on the 555 timer to the power supply bus.

Connect pin 1 on the 555 timer to GND.

Connect pin 2 on the 555 timer to pin 6 on the 555 timer.

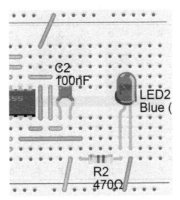

Connect pin 4 on the 555 timer to the power bus on the breadboard.

Connect pin 3 on the 555 timer to the LED's limiting resistor (resistor R2, the 470 ohm one).

Place the LED on the breadboard and connect its cathode (the short leg) to GND.

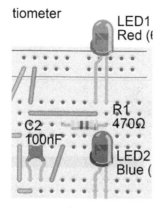

Place another limiting resistor, R1, for the additional LED.

Place the second LED on the board and connect its anode to the positive power rail.

The 100 kOhm potentiometer changes the LED blink frequency anywhere from every 2 seconds to 10,000 times per second.

Changing to a 10 MOhm potentiometer will give a frequency range from 2 times per second to 100 times a second.

If you want to just have one flashing light, remove one LED and the 470 ohm paired resistor from the circuit. After that, the remaining LED will be flashing.

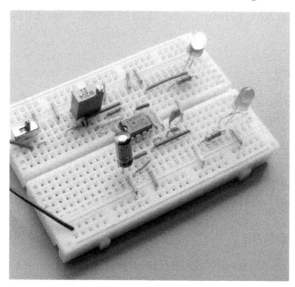

Red LED flashing

Blue LED flashing

Takeaways

- Knowing the relationship between the value of resistors and the speed of the 555 output pulse.

- Knowing how to synchronize two LEDs in this way without a microcontroller.

Things to Try

- As the schematic shows, you can remove one of the LEDs and resistors to create a single flashing light.

- Depending on the voltage source you use and the forward voltage drop on the LEDs, you could put more LEDs in parallel with the single flasher to create a brighter light. If you stick with the 9 V battery, you could only add one more LED before you start experiencing errors or dimmer lights.

Troubleshooting

- The LED flasher circuit has many wires connecting its pins in various ways. It can get confusing. In the breadboard drawing, I used a lot of short wires to make it flat-looking, neater in the end. However, being off by one spot will clobber the operation.

CHAPTER 4

Traffic Light Controller

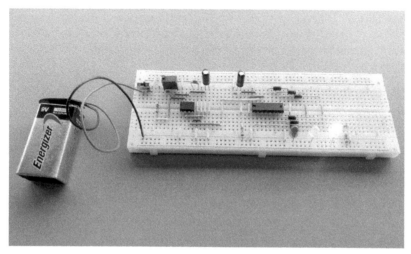

Traffic light circuit giving the OK-to-drive signal

Traffic lights—known as traffic control signals—are everywhere. There are over 300,000 in the United States and millions around the planet.

The decision to use red, amber, and green for automobile signals was an easy one—the same colors, with the same meanings, had been in use for railroad signals since the middle of the 19th century.

They tell drivers which lanes of traffic have the right of way, indicating when cars should go, slow down, and stop by lighting up indicators of universally agreed upon colors (red, amber, and green). Though they seem like complex systems, a 555 timer is a perfect device to use as the basis of a simple traffic light.

Notice we said that the 555 is "the basis" of a traffic light. A 555 is a versatile component, but sometimes you want to build a gadget to perform actions that are outside the scope of a timer, such as counting from 0 to 9. Fortunately, other integrated circuits can do just that, and they can be triggered, or controlled, by 555 timers. For this project, we're going to use the 555 timer to control another circuit. While you probably could make a traffic light system from a hodgepodge of old electromechanical circuits, this is a good point to introduce an important new integrated circuit that adds functionality to the 555 timer.

Meet the 4017 IC, also known as the CD4017, IC 4017, Counter, and Divider. It's a 16-pin CMOS Decade Counter Integrated Circuit that will make this and the next project much simpler to build. Like all ICs (including the 555 itself), the 4017 is nothing but a large collection of discrete electrical components arranged in different circuits, placed into a small silicon chip.

If you're curious what the CD stands for in CD4017, it's sort of simple. It represents the first two words in the description of the IC, *CMOS*-Decade counter/divider. CMOS stands for *complementary-symmetry metal-oxide-semiconductor*, which is a way to make low-power transistors.

Inside the 4017 IC is a series of inverse feedback ring counter circuits, referred to as *Johnson Counters*. The ring counters are a collection of two opposing transistors and some resistors, which together act almost like a single bit computer memory. Without any electricity, the counter is set to 0. When an electrical pulse comes in, the counter is set to 1. When another electrical pulse comes in, the counter pulses the *next* counter in the circuit, setting it to 1, and sets itself back to 0. There are ten of these ring counter circuits, and they cascade from one to the other. When the tenth counter is triggered, it triggers the first counter in the series and the cycle starts all over again.

The counter circuit has two stable states, 0 and 1, and with each electrical pulse it flip-flops between the two states. In the poetic voice of the electrical engineer, such a circuit is called a *flip-flop*. Zooming in to the electrical components inside the 4017 is a great way to show how dense the IC is, which is why we use it instead of using full-sized transistors and resistors to make the circuit.

Now, think of how a traffic light works. Picture what it might be like if the green and red lights of a traffic signal were connected to the output pins of a 4017. Suppose the green light is connected to pins 0, 2, 4, 6, and 8 and the red light is connected to pins 1, 3, 5, 7, and 9. When the 4017 receives its first pulse, the green light comes on. When the second pulse comes in, the green light shuts off and the red light comes on. When the third pulse comes in, the green

In the computing world, a counting sequence usually starts with zero.

light comes on, the red light shuts off, and so on, and so on. A traffic light is, ultimately, nothing more than a timed sequence of lights.

Just as the 555 timer can't exactly perform what the 4017 can, the 4017 cannot operate without an input signal. In this project, the 555 timer will provide that signal. The 555 will be placed into astable mode, producing a square wave with an exact frequency of pulses. Those pulses will control the 4017's LEDs, producing the traffic light sequence.

Parts

1x Counter ICs, CD4017	Newark part number 60K5099	
555 Timer IC	Newark part number 58K8943	
6x Schottky Diode, 1N4004	Newark part number 12T2303	
2x Red LED	Newark part number 52K5254	
2x Yellow LED	Newark part number 97K4048	
2x Green LED	Newark part number 97K4041	
330 ohm Resistor	Newark part number 28X2253	
470 ohm Resistor	Jameco part number 690785	Newark part number 58K5055
1 kOhm Resistors	Jameco part number 690865	Newark part number 38K0327
10 kOhm Resistor	Jameco part number 691104	Newark part number 58K3886
20 kOhm Resistor	Newark part number 58K5026	
100 kOhm Potentiometer	Newark part number 62J1426	
1 µF Ceramic Capacitor	Newark part number 97M4165	
2x 10 µF Electrolytic Capacitor	Newark part number 63K2677	

9 V Battery Holder/Strap	Newark part number 31C0662 or 59K0356
9 V Battery	Newark part number 81F157
2x Slider Switch SPDT	Newark part number 10X9279
1x Full-sized Breadboard	Newark part number 99W1760

The Official Schematic

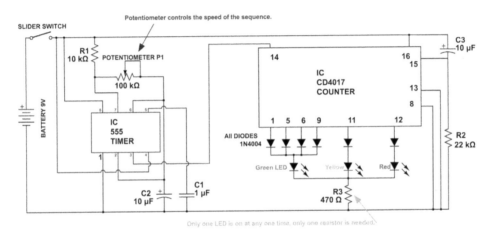

The schematic shows the new IC, the 4017. Keep in mind the pins of the 4017 are not labeled in order like the 555 timer. They're labeled for convenience and to show the LEDs in a line.

A traffic light project showcases the functionality of a 4017 perfectly. It's a clever use of a straightforward signal progression through the output pins. One output after another of the 4017 is set high, lighting up green, amber, and red LEDs, and then the whole process starts over.

All ten output pins of the 4017 are labeled, from the first output Q0 to the final output Q9. Unfortunately, the pin numbers that correspond to the outputs are not arranged in a line:

Q0 = Pin 3

Q1 = Pin 2

Q3 = Pin 7

Q4 = Pin 10

Q5 = Pin 1

Q6 = Pin 5

Q7 = Pin 6

Q8 = Pin 9

Q9 = Pin 11

The mixup of numbers is due to the internal logic of the 4017 IC. When the IC was designed, it was just more convenient to lay the pins out this way, or perhaps it was just too cramped inside the 4017 to do it any other way. Whatever the reason, this *pin-out*, as it's called, is standard for this particular IC.

Note that the first 5 pins of the sequence Q0 to Q4 are not being used in the schematic or breadboard view. There's a reason, and it has to do with our strive for a simpler circuit. In the 4017, pin 12 actually outputs a logic high signal for the first five counts the chip receives. So, we can light up an LED for five counts with a single pin.

The sequence is as follows:

The first five outputs, Q0 to Q4, will energize pin 12, lighting up the red LED.

Q5, on pin 1, is the green LED.

Q6, pin 5, is also green.

Q7, pin 6, is green.

Q8, pin 9, is green.

Q9, on pin 11, is connected to the yellow LED.

So this particular traffic light is red for 5 counts, green for 4 counts and yellow for one count.

By using pin 12, we're saving components and in general simplifying the circuit. We don't *have* to do it this way to make a traffic light, but it is pretty handy. In the circuit as built, the first five pulses sent by the 555 timer to the 4017 will light up the red LED on pin 12. By wiring the project this way, we're severely reducing complexity.

Notice also that each LED is connected to the 4017 via a diode. The diode blocks electrical signals from being fed back into the 4017 IC from other pins or sources. Feedback could damage the IC, or at least mess with the IC's operation.

Breadboard

In the schematic view, the pin connections are simplified for an easier to follow graphic. However, in the breadboard view shown in the top image on page 37, we see exactly how the components are connected to each other.

Since only one LED should be lit during the traffic light's operation, all LEDs share a single resistor. This, again, is to reduce the component count in the build. Reducing the component count will also be helpful in organizing the breadboard for the project. In this case, all the negative sides of the LEDs are connected together at the bottom power rail, which is a single shared electrical point. Since we have to connect the LEDs at one point, this will help. Then, from there, it goes back up to the 470 ohm resistor and ultimately to ground.

Take a look at that breadboard and imagine what it would look like if we were using pins Q0 through Q4 directly. Try to picture four additional diodes crammed somewhere around that 4017 IC. It's a good thing we're using the pin 12 trick!

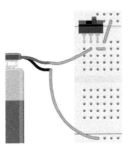

Connect the negative connection on the battery to the GND bus on the breadboard. Place the slider switch on the breadboard and connect the positive connection of the battery to the middle pin on the slider switch. Connect the right pin on the slider switch to the power bus on the breadboard.

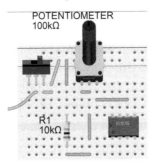

Place the 10 k resistor on the breadboard and connect one lead to the power bus. Place a potentiometer on the breadboard and connect the open lead on

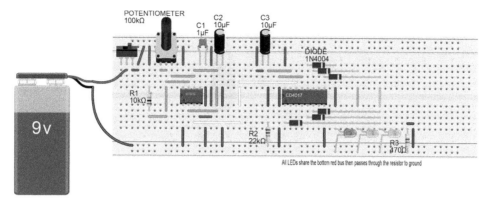

POTENTIOMETER
100kΩ

C1
1µF

C2
10µF

C3
10µF

DIODE
1N4004

R1
10kΩ

R2
22kΩ

R3
470Ω

9v

All LEDs share the bottom red bus then passes through the resistor to ground

The breadboard view shows just how confusing a circuit can get in reality.

the 10 k resistor to the middle pin on the potentiometer. Connect the middle pin and left pin on the potentiometer. Place the 555 timer on the breadboard and connect the pin 6 to the right pin on the potentiometer.

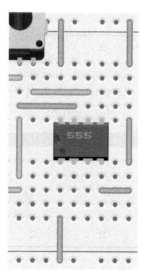

Connect pin 7 on the 555 timer IC to the middle pin on the potentiometer. Connect pin 8 on the 555 timer to the power bus on the breadboard. Connect pin 4 on the 555 timer to the power bus on the breadboard. Connect pin 1 on the 555 timer to the GND connection on the breadboard.

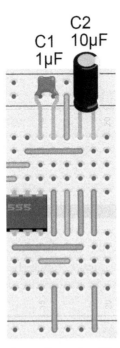

Place a 1 µF capacitor on the breadboard with one lead connecting to pin 5 on the 555 timer IC. Connect the open lead on 1 µF capacitor to GND on the breadboard. Place a 10 µF capacitor on the breadboard and connect its anode to pin 6 on the 555 timer IC. Connect pin 2 of the 555 timer to the anode on the 10 µF capacitor. Connect the cathode of the 10 µF capacitor to GND connection on the breadboard.

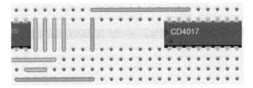

Place the 4017 counter IC on the breadboard and connect pin 14 of the counter IC to pin 3 of 555 timer IC.

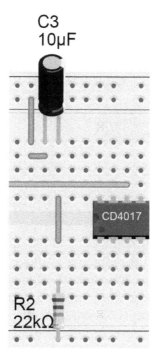

Place the 10 µF capacitor on the breadboard with the anode connecting to the power bus on the breadboard. Place the 22 kOhm resistor with one lead connecting to the cathode of 10 µF capacitor and one lead connecting to GND on the breadboard.

Connect pin 15 of the counter IC to the cathode on 10 µF capacitor. Connect pin 16 of the counter IC to power on the breadboard. Connect pin 13 of the counter IC to GND on the breadboard. Connect pin 8 of the counter IC to GND on the breadboard.

Place a diode with the anode connecting to pin 12 on counter IC. Place a red LED with its anode connecting to the pin 12 diode and its cathode connecting to the bottom red bus on the breadboard—this will become a common LED bus. Place a diode with its anode connecting to pin 11 on counter IC. Place a yellow LED with its anode connecting to pin 11's diode and its cathode connecting to the LED bus.

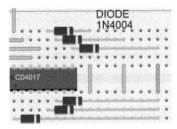

Place a diode with its anode connecting to pin 9 of the counter IC and its cathode connecting to a node that will become a common diode node. Place a diode with its anode connecting to pin 6 of the counter IC and its cathode connecting to the node that will become a common diode node. Place another diode with its anode connecting to pin 5 of the counter IC and its cathode connecting to the common diode node. Place yet another diode with its anode connecting to pin 1 of the counter IC and cathode connecting to the common diode node.

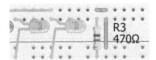

Connect a green LED with its anode connecting to the common node of four diodes and its cathode connecting to the LED bus. Place the 470 ohm resistor so that it connects from the LED bus to the GND connection.

Once the circuit is powered up, you'll have a simple traffic light. The duration of the signals is controlled by the potentiometer. Adjusting the potentiometer clockwise should lengthen the duration of the lights. Counterclockwise should speed up the lights. The next time you see a traffic light for real, imagine that it just might be counting through output pins of a 4017 IC.

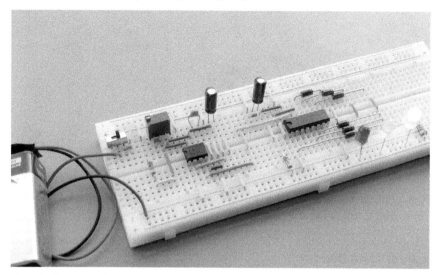

Here is the circuit, again giving you the OK to drive.

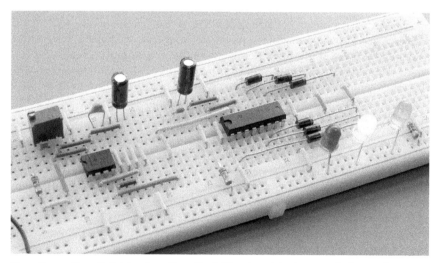

Now the yellow light is signaling that it's about to turn red. Since this LED is on only one output pin of the 4017, it will have the shortest duration of the three colors.

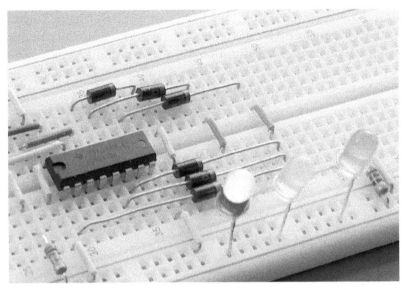

Red light, signaling a stop. This is the longest light. It is lit up for 5 counts, compared to the 4 for green and 1 for yellow.

Counter...and Divider

Back in paragraph three of this chapter, we said that the 4017 is also a divider IC. How can something that counts from 0 to 9 also be a divider?

Suppose you have a circuit consisting of a 555, a 4017, and another 555. Call the two 555s A and B. 555A puts out ten pulses every second, or 10 Hertz. It feeds those pulses into the input of the 4017. Now suppose that *only* output 1 of the 4017 is fed into the input of 555B and all the other outputs of the 4017 are disconnected. When 555A sends a pulse to the 4017, only one pulse in every 10 is sent to 555B. As far as 555B is concerned, it is receiving one-tenth as many pulses as 555A is putting out. We have divided a 10 Hz signal into a 1 Hz signal.

You can mix and match the outputs of the 4017 any way you like. Suppose you connect every other 4017 output to 555B. That would send every other pulse to the timer, effectively dividing a 10 Hz signal in half into a 5 Hz signal. Or you can connect pins 1 and 6, thereby dividing the 10 Hz signal by 5, resulting in a 2 Hz signal. Unfortunately, there aren't any other whole number divisors of 10 other than 10, 2, and 5. You can try connecting every third output pin of the 4017, but that wouldn't give you a smooth 1 pulse in 3—since 10 can't be evenly divided by 3, there's always a little hiccup as the remainder gets carried to the fourth pulse.

Further Ideas

- You might want to move the LEDs to another place, perhaps away from the control circuit. You'll have to make some longer leads to attach to each LED. Connect the leads to the control board in the same positions for each respective LED.

- Suppose this traffic signal were controlling a north-south road. What would the traffic signal for an intersecting east-west road look like? At first thought, you might believe that the lights would be the opposite: when one signal is green, the other would be red. But would both signals be amber at the same time? How would you adjust for that?

One thing you have to take into account is the voltage of the circuit. If, for example, you power a green LED at the same time as the red LED, then you would need to change the resistor to accommodate for the added voltage drop. A lower value resistor, like 220 ohms, might help keep the LEDs bright.

- With enough LEDs, can you make a four-way traffic signal?

- The first four pins on the 4017 will have the north/south lights green, and east/west's lights red.

- The fifth pin makes the north/south signal yellow, keeping east/west red.

- The sixth to the ninth pins will be a separate circuit, turning the east/west lights green, and the north/south lights red.

- The tenth pin controls east and west's yellow light, while still powering north and south's red LED.

- Then the sequence starts over.

CHAPTER 5

LED Chaser

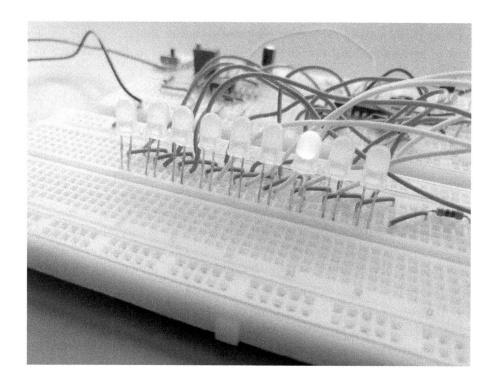

Chaser lights—add instant cool to anything.

Time to truly "own the light." In this chapter, we're going to build chaser lights. You know what they are: a series of lights that illuminate one after the other, either in one direction or back and forth.

The eerie effect of a scanning light was first made famous in *The Day the Earth Stood Still* (1951) when the giant extraterrestrial robot Gort came to life after landing on Earth. His scanning light, placed where his eyes should be, concealed a disruptor beam that could destroy weapons without harming people.

Many years later we saw chaser lights again in helmets of the sinister Cylons in *Battlestar Galactica* (1978–1979). But it probably was the TV show *Knight Rider* (1982–1986) that shot the chaser light effect skyward in terms of popularity. The Knight Rider's car, a black Pontiac Trans Am named K.I.T.T, sported the chaser light effect on the front of the grill. It was the coolest thing around, at the time.

The chasing lights effect has been used in TV shows, movies, and stage performances ever since. Ultimately, they're simply a row of lights that light up back and forth, but they look amazing.

> Both Battlestar Galactica and Knight Rider were produced by a man named Glen A. Larson, who really liked the chasing lights effect. For this reason, sometimes you'll see the effect called a "Larson Scanner," in his honor.

To make this project work, in addition to a 555 timer, we're going to again need that integrated circuit called the CD4017. The 4017, like any IC, has been built to do just one thing: in this case, it counts from 1 to 10.

When the 4017 receives a pulse of electricity into its input pin, it energizes output pin 1. If an LED is connected to pin 1, that LED will light up. When the 4017 receives a second pulse of electricity into its input pin, it shuts off output pin 1 and powers up output pin 2. LED number 1 will go dark, and LED number 2 will light up. On a third pulse of electricity, the 4017 shuts off pin 2 and lights up pin 3, and so on up to 10. When the 4017 receives an eleventh pulse, it shuts off pin 10 and starts all over again with output pin 1.

The circuit for the chaser lights should be pretty easy to picture in your mind. A 555 sends a pulse to a 4017, which lights up a row of LEDs one after another. In this project, we'll re-create that effect on a breadboard. I wouldn't be at all surprised if you tried to attach this project to something you wear, ride, or use. Chaser lights are just so appealing, truly lit.

Parts

2x Counter ICs, CD4017	Newark part number 60K5099
555 Timer IC	Newark part number 58K8943
15x Schottky Diode, 1N4004	Newark part number 12T2303
11x LEDs of your choice.	Newark part number 52K5254

2x 2N2222 Transistor	Newark part number 87K2254
330 Ω Resistor	Newark part number 28X2253
1 kΩ Resistor	Newark part number 38K0327
2.7 kΩ Resistor	Newark part number 59K8360
2x 4.7 kΩ Resistor	Newark part number 38K0304
10 kΩ Resistor	Newark part number 38K0328
100 kΩ Potentiometer	Digikey part number CF14JT100KCT-ND
1 µF Ceramic Capacitor	Newark part number 97M4165
10 µF Electrolytic Capacitor	Newark part number 63K2677
9 V Battery Holder/Strap	Newark part number 31C0662 or 59K0356
9 V Battery	Newark part number 81F157
2x Slider switch SPDT	Newark part number 10X9279
2x Full-sized Breadboard	Newark part number 99W1760

The Official Schematic

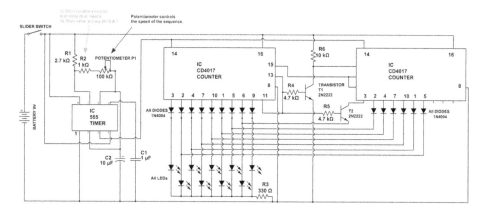

The schematic shows the new ICs in use. Keep in mind the pins of the CD4017 are not labeled in order like the 555 timer. They're labeled for convenience and to keep the LEDs in a line.

In this circuit, two 4017 counter ICs are used to drive the same set of 11 LEDs. The 555 timer, in astable mode, sends a clock signal, or pulse, to the first counter IC. The counter then goes through its light-up sequence defined by the pulse's speed, or frequency. When the first counter is finished, it triggers the second counter. The second IC drives the LEDs to light up one after the other, but in reverse. Notice the second 4017 IC is wired to the LEDs in the reverse order compared to the first 4017 IC.

Another new component is being introduced in this circuit, a diode. Diodes are critical for this project to work correctly. Wait, aren't LEDs light-emitting diodes? Are they the same? Kind of, but no.

A regular diode is an electrical component with two leads, like a resistor. However, this component only lets current flow in one direction. It has a low resistance in the direction of current flows and acts like a huge resistor in the opposite direction. A huge resistance often looks like a wire connected to nothing inside a circuit. A diode is an effective way to keep signals and voltages from feeding back into parts of your projects.

Can an LED also block voltage in the reverse direction? Yes, but nothing like a regular diode.

The first 4017 will light up the LEDs in sequence, from left to right. When the 4017 gets to pin 11, the last output on that IC, it will stop counting on that IC. Pin 11 is connected to pin 13, which is called a *clock enable*. When it's set to LOW, the count stops.

This signal from pin 11 also triggers a transistor (T1) to apply a logic low signal to the second IC's *reset* pin, pin 15. This enables counting on the second 4017 IC. The second transistor (T2) is connected to the next LED in the reverse series, which starts the reverse lighting effect.

Output 7, pin 6, on the second IC is connected to the second IC's clock enable, disabling the chip's count. At the same time, energizing this pin re-enables the first 4017 IC's reset pin, pin 15.

I know it's a brute force method to light the LEDs in reverse, but it works, and it works well.

The circuit will bounce back and forth between the two 4017 ICs as long as it's powered. The Cylon-Knight Rider-Gort effect is complete!

Breadboard

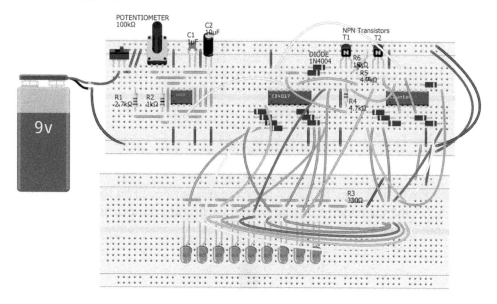

The breadboard view shows just how confusing a circuit can get in reality.

You could make the entire project fit on one large breadboard, but that'll be one cramped board! I suggest making the LED line on a separate board.

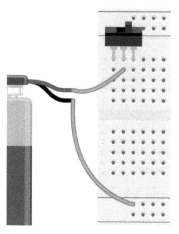

Connect the negative connection of the battery to the GND bus on the breadboard. Place the slider switch on the breadboard and connect the positive connection of the battery to the middle pin.

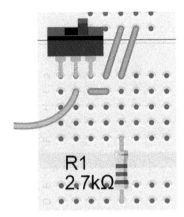

Connect the right pin on the slider switch to the power rail on the breadboard. Place a 2.7 kOhm resistor on the breadboard and connect one lead to the power rail on the breadboard.

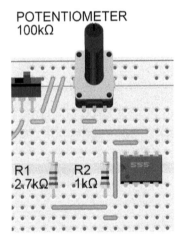

Place a potentiometer on the breadboard and connect its left pin to its middle pin.

Place a 1 kOhm resistor on the breadboard with one lead connecting to the right pin on the potentiometer.

Connect the open lead of the 1 kOhm and 2.7 kOhm resistors with a wire.

Place the 555 timer IC on the breadboard and connect pin 7 to the node connecting the 1 kOhm resistor and the 2.7 kOhm resistor.

Connect pin 6 on the 555 timer IC to the middle pin of the potentiometer.

Connect pin 8 on the 555 timer IC to the power bus on the breadboard.

Connect pin 1 on the 555 timer IC to the GND connection on the breadboard.

Place the 1 µF capacitor on the breadboard and connect one lead to pin 5 on 555 timer IC.

Connect the other lead of the 1 µF capacitor to the GND connection on the breadboard.

Connect pin 4 on 555 timer IC to the power bus on the breadboard.

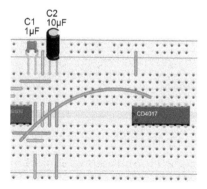

Place the 10 µF capacitor on board and connect the anode to pin 6 on the 555 timer IC.

Connect pin 2 on the 555 timer IC to the anode of the 10 µF capacitor.

Connect a cathode of 10 µF capacitor to GND connection on the breadboard.

Place counter IC 1 on the breadboard and connect pin 14 to pin 3 on the 555 timer IC.

Connect pin 16 of the counter IC 1 to the power bus on the breadboard.

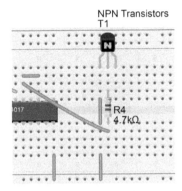

Connect pin 13 to pin 11 on the counter IC 1.

Connect pin 8 of the counter IC 1 to the GND connection on the breadboard.

Place a 4.7 kOhm resistor on the breadboard and connect one lead to pin 13 of the counter IC 1.

Place transistor T1 on the breadboard and connect the base of the transistor to the open lead on 4.7 kOhm resistor.

Connect the emitter of transistor T1 to the GND connection on the breadboard.

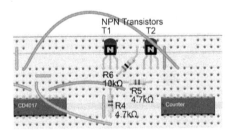

Place a 10 kOhm resistor on the breadboard with one lead connecting to the power bus and one lead connecting to the transistor T1 collector lead.

Place a 4.7 kOhm resistor (R5) on the breadboard and connect one lead to pin 11 on counter IC 1.

Place transistor T2 and connect the base to the open lead of 4.7 kOhm resistor R5.

Place counter IC 2 on the breadboard and connect pin 16 to the power bus on the breadboard.

Connect pin 14 of the counter IC 2 to pin 14 of the counter IC 1.

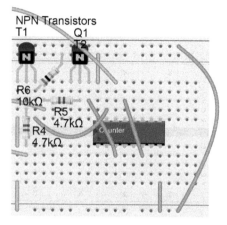

Connect pin 15 of the counter IC 2 to collector lead on transistor T1.

Connect pin 3 of the counter IC 2 to the collector lead on transistor T2.

Connect pin 6 to pin 13 on the counter IC 2.

Connect pin 8 of the counter IC 2 to GND connection on the breadboard.

Connect the two GND bus rails on the top breadboard.

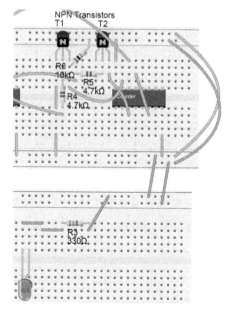

Connect the two power bus rails on the top breadboard.

Connect the power bus between the breadboards.

Connect the GND bus between the breadboards.

Place a 330 ohm resistor on the breadboard and connect one lead to the GND bus.

Place LED 9 on the breadboard and connect the cathode to the open lead on 330 ohm resistor.

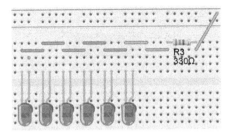

Place LED 8 on the breadboard and connect its cathode to LED 9's cathode.

Place LED 7 on the breadboard and connect its cathode to LED 8's cathode.

Place LED 6 on the breadboard and connect its cathode to LED 7's cathode.

Place LED 5 on the breadboard and connect its cathode to LED 6's cathode.

Place LED 4 on the breadboard and connect its cathode to LED 5's cathode.

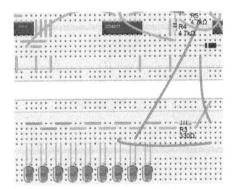

Place LED 3 on the breadboard and connect its cathode to LED 4's cathode.

Place LED 2 on the breadboard and connect its cathode to LED 3's cathode.

Place LED 1 on the breadboard and connect its cathode to LED 2's cathode.

Connect the anode of LED 8 to the emitter of transistor T2.

Place a diode with the anode connecting to pin 2 of the counter IC 2, and connect the cathode to the LED 7 anode.

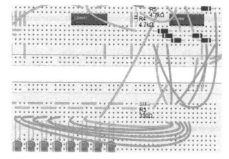

Place a diode with the anode connecting to pin 4 of the counter IC 2, and connect the cathode to LED 6's anode.

Place a diode with the anode connecting to pin 7 of the counter IC 2, and connect the cathode to the LED 5 anode.

Place a diode with the anode connecting to pin 10 of the counter IC 2, and connect the cathode to the LED 4 anode.

Place a diode with the anode connecting to pin 1 of the counter IC 2, and connect the cathode to LED 3 anode.

Place a diode with the anode connecting to pin 5 of the counter IC 2, and connect the cathode to the LED 2 anode.

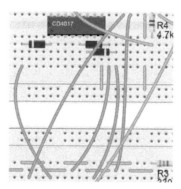

Place a diode with the anode connecting to pin 9 of the counter IC 1, and connect the cathode to the LED 9 anode.

Place a diode with the anode connecting to pin 6 of the counter IC 1, and connect the cathode to the LED 8 anode.

Place a diode with the anode connecting to pin 5 of the counter IC 1, and connect the cathode to the LED 7 anode.

Place a diode with the anode connecting to pin 1 of the counter IC 1, and connect the cathode to LED 6 anode.

Place a diode with the anode connecting to pin 10 of the counter IC 1, and connect the cathode to the LED 5 anode.

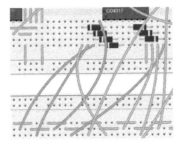

Place a diode with the anode connecting to pin 7 of the counter IC 1, and connect the cathode to the LED 4 anode.

Place a diode with the anode connecting to pin 4 of the counter IC 1, and connect the cathode to the LED 3 anode.

Place a diode with the anode connecting to pin 2 of the counter IC 1, and connect the cathode to the LED 2 anode.

Place a diode with the anode connecting to pin 3 of the counter IC 1, and connect the cathode to the LED 1 anode. All done!

Did I say you could put this on something you wear? You can. Consider this ideal place—put the controller circuit in a jacket pocket, then extend the LED's lead wires to mount wherever you want, becoming the Knight Rider.

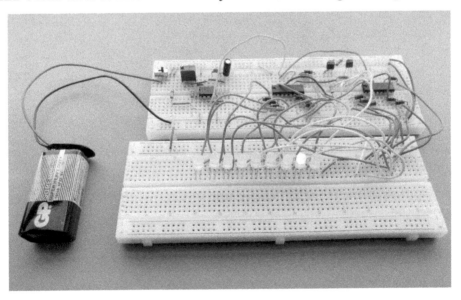

Another view of the LED chaser breadboard—yes, the wires are messy. Don't fear having a messy breadboard, as long as you know what's happening in your circuit. Feel free to tighten up the layout.

Takeaways

- Using CD4017 counter ICs to control LEDs.

- Creating light effects. This is a lighting effect that would be hard to accomplish without the counter ICs. A chaser light circuit is one of those lighting effects that are universally appealing, and it's good to know how to build one. Imagine this circuit with sound output too. It's possible.

- Do yourself a favor and check out those shows and films I mention in the beginning: the original *The Day the Earth Stood Still* (1951), the TV series *Battlestar Galactica* (1978–1979), and especially the TV series *Knight Rider* (1982–1986). You won't regret it.

Things to Try

- Place the controller circuit in a jacket pocket, then extend the LED's lead wires to mount wherever you want, such as down the sleeve of a jacket or across the back. Becoming the Knight Rider—you're immediately cooler.

- The pins from the second 4017 IC don't have to be connected in reverse order. You can hook up twenty LEDs that would light up one-by-one in a single direction and then reset to the beginning.

- Similarly, you could add more 4017 ICs, repeating the same schematic setup but chaining off each other. Then the last one can feedback to the first. There is, technically, no limit to the length of this chain.

- You could also connect the same 555 timer to several sets of 4017 circuits to create more LEDs bouncing back and forth. I suppose that could also be done by putting LEDs in parallel.

- Try making a countdown, ready-set-go, light. By slowing down the 555 timer, the speed at which the LEDs light up will be slow. You can place a green GO LED at the end of the sequence. Then also disconnect pin 6 on the second 4017 and the sequence will only work once before needing to be reset. But there's no re-do in ready-set-go, right?

Troubleshooting

- The LED Chaser has A LOT OF WIRES! Like in the breadboard drawing, if you can get different color wires for every LED, this will help make troubleshooting go way faster.

- Another point to consider is diode orientation. There are fifteen diodes here; make sure they point the correct way or else some LEDs won't light up.

- Again, short wires can look like they are in the right spot, but look closer.

CHAPTER 6

LED Brightness Dimmer

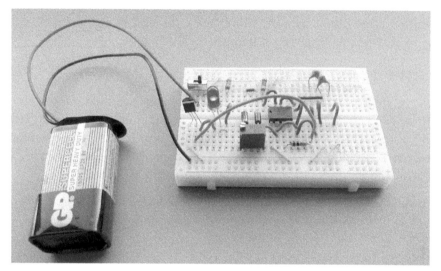

Learning how to dim an LED is as important as learning how to turn one on.

You might be wondering: "Are LEDs dimmable? Can I just use a potentiometer?" Sure you can, but with one caveat to that solution: when you lower the amount of electricity going to the LED using a resistor or potentiometer (variable resistor), all the excess energy gets converted to heat. Let's say you have a 9 volt battery in the setup, and you want to dim a 2.1 volt LED. That means the potentiometer must dissipate nearly 7 volts of energy. Unless that potentiometer is huge, there's a good chance it's going to get really, really hot.

Deliberately designing a circuit that you know is going to heat up is dangerous for many reasons, so let's defeat the heat. What you need is a way to reduce the amount of energy going to an LED without building up waste heat. The solution is something called *pulse width modulation*, also known as PWM.

In electronics, a *pulse* is a discrete burst of energy with a beginning, middle, and an end. When the pulse ramps up to peak voltage, the signal is said to be *high*. When the pulse reduces to zero, the signal is said to be *low*. The middle portion of the pulse, also known as the *width*, can last any amount of time: a fraction of a second, a few seconds, a minute, or an hour or more. The distance from one pulse to the next pulse is called a *duty cycle*. If the signal is at a peak voltage 100% of the time, it's called 100% duty cycle. Peak voltage only half the time is a 50% duty cycle, and so on.

To dim an LED, we *modulate*, or change the signal so that it's at peak voltage for only half the time and absolutely off for the other half of the time—a 50% duty cycle. This accomplishes our goal of reducing the amount of energy going to the LED, except that instead of reducing the voltage by half, we're reducing the time the LED is lit by half. Either way, the end result is that the LED appears only half as bright as it should.

Persistence of Vision

Obviously, if the LED's duty cycle is too long, you won't see a dimming effect, just a light blinking on and off. But because of the way the human eyes and nervous system work, an LED blinking at hundreds or thousands of times per second will be perceived as a dim LED. It's almost as if the brain averages an image of a fully lit LED and a fully dark LED into a mental image of a 50% lit LED.

By using pulse width modulation to dim an LED, no electrical power is wasted as heat, as it would be if we used a potentiometer.

Parts

555 Timer IC	Newark part number 58K8943	Jameco part number 27423
NPN Transistor, 2N3904	Newark part number 83C3116	Jameco part number 178597
LEDs of your choice. I suggest red.	Newark part number 52K5254	Jameco part number 206519
2x Diode, 1N4148	Newark part number 95W3791	Jameco part number 36038
220 Ω Resistor	Newark part number 38K0351	Jameco part number 690700
2x 1 kΩ Resistor	Newark part number 38K0327	Jameco part number 690865

100 kΩ Potentiometer	Digikey part number CF14JT100KCT-ND	Jameco part number 853599
0.01 μF (10 nF) Ceramic Capacitor	Newark part number 46P6665	Jameco part number 15229
0.1 μF (100 nF) Ceramic Capacitor	Newark part number 46P6667	Jameco part number 25523
9 V Battery Holder/Strap	Newark part number 31C0662 or 59K0356	Jameco part number 101470
9 V Battery	Newark part number 81F157	Jameco part number 27423
Slider Switch SPDT	Newark part number 10X9279	Jameco part number 2192384
Half-size Breadboard	Newark part number 99W1759	Jameco part number 2157693

The Official Schematic.

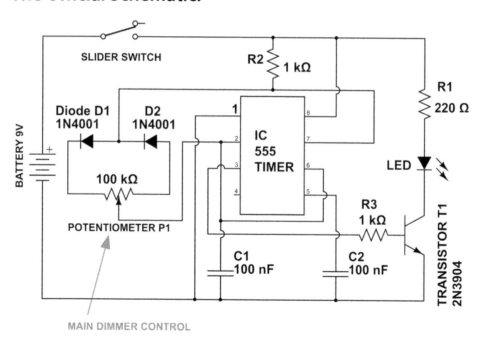

Pay special attention to how to orient the diodes in the circuit. This is an elaborate circuit, so we're going to use a full breadboard.

This circuit can be used as an accessory in many of the other projects. You could use it to control the speed of a small motor like the kind you find inside old broken toys: just hook up the motor where the LED and resistor sit above the transistor.

Breadboard View

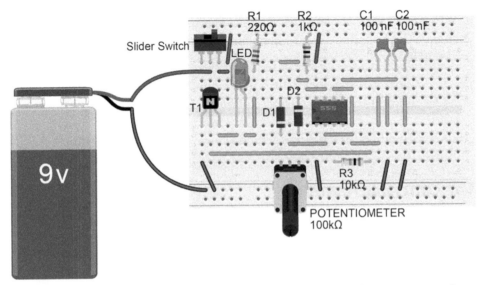

This is the breadboard view. Again, pay close attention to the orientation of the diodes.

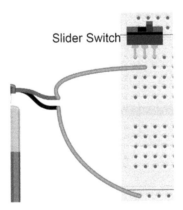

Place switch on the breadboard and connect positive lead on the battery to the middle pin on switch (common node).

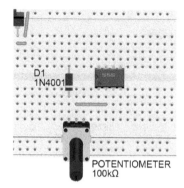

Connect one lead of the switch to the power rail on the breadboard.

Place the 555 timer on the breadboard.

Place the potentiometer on the breadboard and connect the wiper (middle pin) to pin 2 on the 555 timer IC.

Place one diode with the cathode connecting to the left pin on the potentiometer.

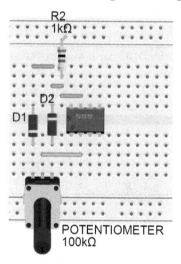

Place one diode with anode connecting to the rightmost pin on the potentiometer.

Connect two diodes on the breadboard with a wire to connect cathode to the anode for the open leads.

Place the 1 kOhm resistor on the breadboard with one lead connecting to the power rail. The additional lead will connect to pin 7 on the 555 timer IC.

Connect the open nodes of the two diodes to pin 7 and the 1 kOhm resistor that connects to VCC.

Place 220 ohm current-limiting resistor on board and connect one lead to power.

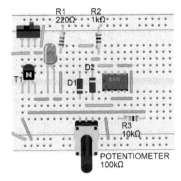

Place LED on the breadboard and connect to 220 ohm resistor's open lead.

Place the NPN transistor on the breadboard and connect the collector to the cathode on the LED.

Connect the emitter of the NPN transistor to GND on the breadboard.

Connect pin 3 on 555 timer IC to the base on the NPN transistor using a 1 kOhm resistor.

Connect pin 6 on the 555 timer to pin 2 on the 555 timer IC.

Connect pin 1 on the 555 to GND on the breadboard.

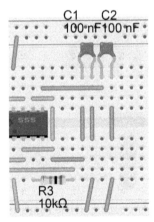

Connect pin 8 on the 555 to power on the breadboard.

Place the 100 nF capacitor on the breadboard and connect to pins 2 and 6 on the 555.

Connect additional lead on the capacitor to GND on the breadboard.

Place the 100 nF capacitor on the breadboard and connect to pin 5 of the 555.

Connect the additional lead on the capacitor to GND on the breadboard.

Note that two of the resistors are connected to the top power rail of the breadboard. This is done to save space. I used a 100 kOhm potentiometer in this design. You can swap it for one with a smaller resistance for a different dimming effect.

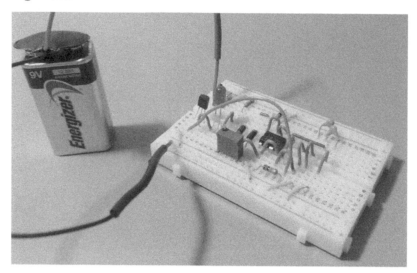

With the potentiometer turned to the middle, the LED is on, but dim.

With the potentiometer turned all the way to the right, the LED is at maximum brightness! With this style potentiometer, it take several turns to reach the end. This one is about fifteen turns.

But Wait!

We started out saying that we shouldn't use a potentiometer to change the brightness of an LED, yet here we are using a potentiometer to change the brightness of an LED. And we're using a much more complicated circuit to do it. Why?

The answer is we're doing the job carefully. Rather than converting useful electricity to waste heat, we're being smart and limiting the amount of time the LED is lit up. It adds a little bit of complexity but results in a safer, longer-lasting circuit. Using the 555 chip is the way to go.

Takeaways

- By modulating, or changing, the amount of electricity going to an LED, we can change its brightness. PWM can also be a way to extend battery life. For example, if you have a 50% PWM signal, it would use half the power of a signal running at 100%.

- This project also adjusts PWM on the fly, meaning as the circuit is operating. This concept can be used to dim a light, of course, but also can be used to slow down a motor.

Things to Try

- You can add another LED and resistor in parallel with the one in the circuit to have a brighter setup.

- You can connect this one to the One-Shot circuit in place of the LED and resistor above the transistor of the One-Shot. What you'll have is a one-shot LED you can dim as you see fit.

Troubleshooting

- The LED Dimmer has a section that is a bit confusing. In the central dual-diode portion, pay attention to orientation and all the short jumper wires. This is most likely the problem area.

CHAPTER 7

555 Timer Silent Alarm

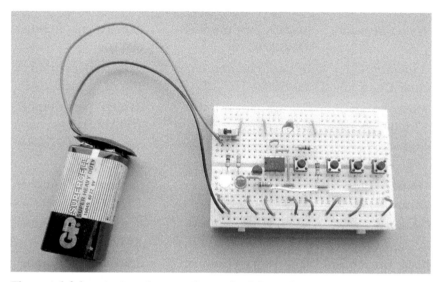

The watchful protector, the guardian...the Silent Alarm

Consider the following scenario:

You suspect someone has been trespassing in a place where you don't want them to. You can't catch them in the act. What do you do? How can you confirm that someone is indeed violating your space?

You need technology. You need a device that alerts *you* to the presence of the trespasser without also alerting the trespasser in the process. You need a Silent Alarm.

Parts

555 Timer IC	Newark part number 58K8943	Jameco part number 27423
A green and red LED	Newark part numbers 97K4041 & 52K5254	Jameco part number 97K4041 & 52K5254
NPN Transistor, 2N2222	Newark part number 58K2047	Jameco part number 178511
470 Ω resistor	Newark part number 58K5055	Jameco part number 690785
1.5 kΩ Resistor	Newark part number 58K5015	Jameco part number 690902
3x 10 kΩ Resistor	Newark part number 38K0328	Jameco part number 691104
0.01 µF (10nF Ceramic Capacitor	Newark part number 46P6665	Jameco part number 15229
4x Momentary Contact Tactile Switch	Newark part number 60M5365	Jameco part number 153252
Slider Switch SPDT	Newark part number 10X9279	Jameco part number 2192384
9 V Battery Holder/Strap	Newark part number 31C0662 or 59K0356	Jameco part number 101470
9 V Battery	Newark part number 81F157	Jameco part number 27423
Half-size Breadboard	Newark part number 99W1759	Jameco part number 2157693

The Official Schematic

Up until now, you've used 555 timers in oscillator mode. Now we're going to put the 555 timer into *bistable* mode and create the simplest form of digital memory, a flip-flop circuit—also known as a Schmitt Trigger, or a hysteresis circuit, as shown on page 71.

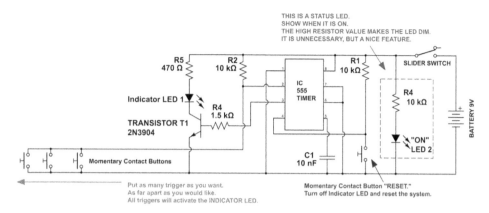

THIS IS A STATUS LED.
SHOW WHEN IT IS ON.
THE HIGH RESISTOR VALUE MAKES THE LED DIM.
IT IS UNNECESSARY, BUT A NICE FEATURE.

Put as many trigger as you want.
As far apart as you would like.
All triggers will activate the INDICATOR LED.

Momentary Contact Button "RESET."
Turn off Indicator LED and reset the system.

The key features of this schematic are the alarm triggers. Make sure to give yourself plenty of wire to place the switches where you think is best.

Bistable mode means just what it says: this particular circuit has two stable states, high and low. This circuit will stay in either of those two states, high or low, until a trigger button is pressed. In this project, we'll directly indicate what state the 555 timer is in with an LED. The high state will have the LED on, and the low state will keep the LED off.

Unlike the previous 555 circuits, bistable mode doesn't require you to create a frequency, or timer, using a resistor and capacitor combo. Bistable mode has no timing. It simply holds one of two states.

In the schematic, if you press any of the momentary contact buttons (or triggers), the 555 timer will go into its high, or on, state. It will stay in this state until the battery runs out or the reset button is pressed.

When the reset button is pressed the 555 timer circuit will go into its low state, turning off the LED, and will wait until a trigger is pressed again. Again, only one of two states can ever happen in this circuit.

A note on the build of the silent alarm:

As explained, once this circuit has been triggered, it will have to be manually reset by you with a button press. Be sure to hide the RESET button somewhere only you can find so your trespasser can't hide their tracks by resetting the circuit themselves. Tuck it into a corner, tape something over it, any way you can think to hide it—hide it.

Breadboard View

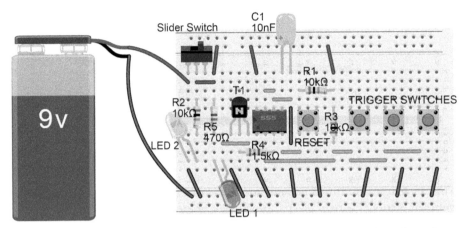

Although in this example all the trigger switches are on this breadboard, in real life you may have to run wires to place the switches where they will be triggered.

First, build a small breadboard of the circuit, so you can see what is happening before you set up a DIY security silent alarm. The three momentary contact switches on the right half of the board are actually three separate triggers for the alarm. You could simply remove them and use long wires to place a contact switch anywhere you want to monitor.

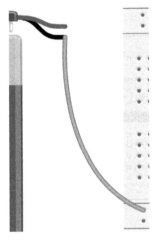

Connect the negative lead of battery to GND terminal on the breadboard.

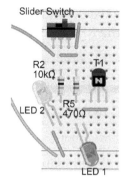

Place switch on the breadboard and connect positive lead of battery to lead on switch.

Connect one end of the switch to the power rail on the breadboard.

Place the 10 kOhm resistor on the breadboard and connect one lead to the power rail.

Place ON LED and connect the cathode to GND on the PCB.

Place the current-limiting resistor on the breadboard and connect one side to the power rail.

Place indicator LED and connect the anode to 470 ohm resistor on the breadboard.

Place NPN transistor and connect collector of transistor to indicator LED anode.

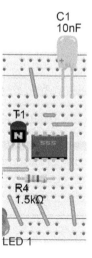

Connect the emitter of NPN transistor to GND on the breadboard.

Place 555 timer IC on the breadboard and connect a 1.5 kOhm resistor from pin 3 on timer to the base of the NPN transistor.

Connect pin 8 on 555 timer IC to power rail on the breadboard.

Connect pin 1 on 555 timer IC to GND on the breadboard.

Connect pin 6 on 555 timer IC to GND on the breadboard.

Connect pins 6 and 7 on the 555 timer.

Place the 10 nF capacitor on the breadboard to connect pin 5 on the 555 timer to GND.

Place reset switch on breadboard and connect one end to pin 4 on 555 timer IC.

Place the 10 kOhm resistor that will connect from switch node to power rail.

Connect the open side of the reset switch to GND on the breadboard.

Place the first trigger switch on the breadboard and connect one side to pin 2 on 555 timer.

Place the 10 kOhm resistor on the breadboard and connect one side to the pin 2 555 timer IC; the other side connects to the bottom left pin of the first trigger switch on the right of the resistor. The other end of the resistor connects to the positive power rail.

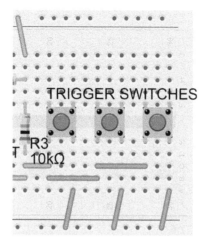

Connect the open side of first switch to GND on the breadboard.

Place second switch and connect one side to the pin 2 555 timer IC node.

Connect the open side of second switch to GND on the breadboard.

Place the final switch and connect one side to the pin 2 555 timer IC node.

Connect the open side of switch to GND on the breadboard.

A number of different types of switches can be wired into this circuit. It just depends on how you want to use it. Momentary contact switches are excellent for such designs, but so are magnetically activated *reed switches*. (A reed switch changes from open to closed or vice versa when in the presence of a magnetic field.)

Tape the magnetic sensor and its magnet inside the lid of a box or trunk. Run the two wires from the reed switch back to where you hid your circuit. Now, when you want to know if someone opened the box/trunk/dresser drawer, check the status lights on your circuit.

You could use a green LED to indicate that all is good and a red LED to indicate that somebody has triggered your expertly concealed alarm circuit.

You may not be able to stop them, but at least you'll know if someone was there.

Silent alarm triggered!

Takeaways

- What happens if you use several different inputs to activate a circuit? Cascading several switches off of one switch still only looks like a single switch from the 555 timer's perspective. You could have a thousand triggers if you wanted.

- Try relocating the trigger inputs. The entire world is your canvas, so don't just confine yourself to a breadboard.

Things to Try

- The Status LED isn't necessary; if you remove that, the circuit gets small and less noticeable.

- As I said in the schematic, there's no end to the number of parallel buttons you can add. You can place more buttons.

- How far can you wire out a button? Miles or kilometers away, and it will still work, technically. However, wires can act as antennas; the longer the wire, the more wireless signals I could pick up. So electrical noise could create a false trigger for your circuit. There are ways to shield and protect this remote switch, but at that point, you might want to look into another solution for long-distance button presses. But I think 1000 feet or around 300 meters would be a safe distance.

• This circuit could technically turn on another circuit if the second circuit is wired up in place of the LED and resistor of the first, between the positive voltage of the battery and the collector of the transistor. This position acts like the battery of the other circuit, functionally. But a single 9 V battery can't power too many circuits at the same time. I'd remove the status LED in this case.

Troubleshooting

• The Silent Alarm is an odd one to troubleshoot, especially if you made long wires to place switches somewhere. If you did that, then start there. Those long wires may have pulled out or broken.

• But if you made it like in this book, then take a closer look at the momentary contact buttons—all of them. Make sure they're connected like in the drawings. Even more important is to check if they are on the board itself; those buttons don't exactly plug into breadboards very securely.

CHAPTER 8

The Single-Button Tunable Synth

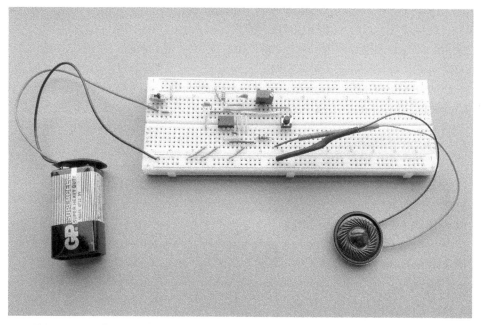

This is one of the simplest synthesizers you can make. Although this single button tunable synth is small, don't underestimate its usefulness.

You've probably heard of a synthesizer (or more likely the more popular shortened term "synth.") Synthesizers use various electronic circuits to create and shape electrical signals that can be turned into sound when played through a speaker. The electrical signals can be created (or *synthesized*) and shaped to sound like other simple musical instruments, such as flutes and brass, or they can be manipulated to create unique sounds. These versatile musical instruments were the basis of most 1980s pop music and continue to be used in almost every song today, whether you realize it or not.

Guess what…you can actually make a synthesizer with a 555 timer! The synth circuit shown in the previous picture is a scaled-down version of the kind used in music, but it isn't any less powerful.

Parts

555 Timer IC	Newark part number 58K8943	Jameco part number 27423
8 Ω Speaker	Newark part number 25R0902	Jameco part number 2234134
2x 1 kΩ Resistor	Newark part number 38K0327	Jameco part number 690865
10 kΩ Resistor	Newark part number 38K0328	Jameco part number 691104
50 kΩ Potentiometer	Newark part number 63M9482	Jameco part number 853791
0.1 μF (100 nF) Ceramic Capacitor	Newark part number 46P6667	Jameco part number 25523
1 μF Ceramic Capacitor	Newark part number 97M4165	Jameco part number 81509
4.7 μF Electrolytic Capacitor	Newark part number 65R3259	Jameco part number 2230109
1x Momentary Contact Tactile Switch	Newark part number 60M5365	Jameco part number 153252
9 V Battery Holder/Strap	Newark part number 31C0662 or 59K0356	Jameco part number 101470
9 V Battery	Newark part number 81F157	Jameco part number 27423
Slider Switch SPDT	Newark part number 10X9279	Jameco part number 2192384
Full-sized Breadboard	Newark part number 99W1760	Jameco part number 2157706

The Official Schematic

The 555 timer synthesizer lets you easily generate a wide range of tones, and a button gives you control over when it plays. Technically, this circuit is more

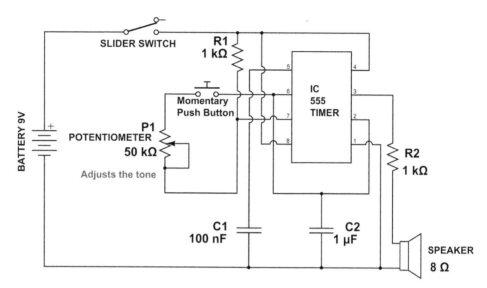

Schematic view of the single-button synthesizer. It's important to note the parts labeled R1 and R2. They determine how this project operates.

like a tone generator than a synthesizer, but it's the foundation of how a full-blown synth will work.

The circuit configuration places the 555 timer into *astable* mode again. As you remember, in astable mode the 555 timer generates a regular on-off pulse at a regular interval. The number of those pulses every second is called the *frequency*. If the frequency of the 555's pulses is in a particular range (about 20 to 20,000 pulses per second), it can be heard through the speaker. In other words, the regularity of the waveform will generate a sound tone.

The potentiometer in the circuit will change the tone drastically if you change the setting as you press and hold the button. You have to operate both components at the same time to hear the music.

What is happening here is a precise controlling of the 555 timer. The resistor labeled R1 and the potentiometer labeled R2 control how long the output pulses are on and off. This is also known as *positive* and *negative time intervals*—the time the pulses are in the on and off state, respectively. The total time between pulses is the positive (on) and the negative (off) time intervals added together. By changing the duration of the off intervals, the potentiometer in this circuit adjusts the frequency of the tones you hear.

Breadboard View

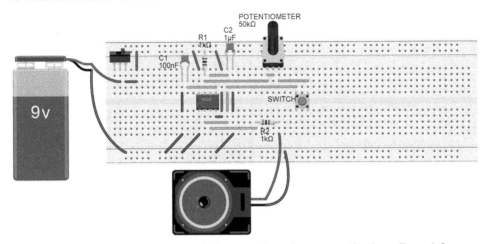

The simple single-button synth leaves a lot of room on the breadboard for expansion.

Since the potentiometer is what changes the tone (and is therefore a component you'll be manipulating a lot), you may want to use one that's easy to turn by hand. However, the one I suggest for this project will give you finer tuning of the potentiometer's value.

The color and exact placement of the wires isn't critical. At this stage, feel free to adjust the layout. Be sure to stick to the wiring of the schematic, though.

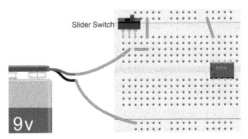

Place the switch on the breadboard and connect the middle pin to the positive lead on the battery.

Connect the negative side of the battery to the GND rail on the breadboard.

Connect an open end of the switch to the power rail on the breadboard.

Place the 555 timer onto the breadboard.

Connect pin 8 on the 555 timer IC to the power rail on the breadboard.

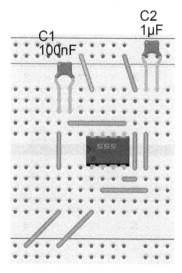

Connect pin 1 on the 555 timer IC to GND on the breadboard.

Place the 100 nF capacitor on the breadboard and connect to pin 5 on the 555 timer IC.

Connect the open lead on 100 nF capacitor to GND connection on the breadboard.

Connect pin 4 on the 555 timer IC to the power rail on the breadboard.

Place the 1 µF capacitor on the breadboard and connect to pin 6 on the 555 timer IC.

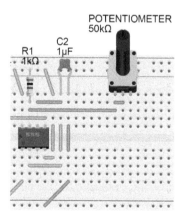

Connect pin 6 on 555 timer IC to 1 µF capacitor node.

Connect the open lead of 1 μF capacitor to GND on the breadboard.

Place the 1 kOhm resistor that connects pin 7 to the power supply rail on the breadboard.

Place potentiometer and connect to pin 7 on the 555 timer IC.

Connect left pin on potentiometer to middle pin (wiper).

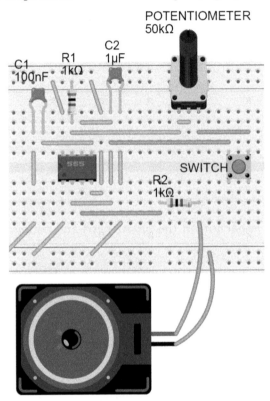

Place momentary switch and connect to pin 2 on 555 timer IC.

Connect the open end of switch to open lead on potentiometer.

Place a 1 kOhm resistor and connect one lead to pin 3 on 555 timer IC.

Place speaker and connect the positive side of the speaker to the open lead on the 1 kOhm resistor.

Connect the negative side of the speaker to the GND rail on the breadboard.

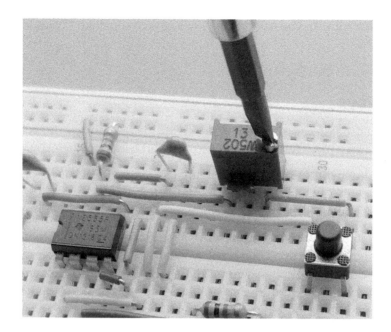

When the circuit is complete, you can play the single-button synth by holding down the button and rotating the potentiometer knob. Turning the knob in one direction makes the frequency (and the tone) go higher. Turning the knob in the other direction makes the frequency and tone go lower.

How long does it take you to learn to play a simple song like "Twinkle, Twinkle, Little Star"?

Takeaways

- You just created a sound generator that demonstrates the relationship of the frequency the 555 timer outputs and the tones heard.

- You adjusted the sound by changing the state of electrical components, specifically, by adjusting the potentiometer's resistance, which changes its relationship with the other components.

- Can you make the one-button synth a better musical "fit"? Most examples of western music don't have notes as low or as high as the one-button synth can produce. Can you adjust the resistors and capacitors of the circuit (using the formulae given earlier in this book) to make the one-button synth fit more naturally into your vocal range?

Things to Try

- If you want a single tone, instead of variable, you can replace the potentiometer with a single resistor of your choice. Not any size resistor can produce a sound you can hear or play through that speaker. I would suggest staying under 30 kΩ. But feel free to experiment on your own.

Troubleshooting

- The single-button synth is probably the easiest to check. There are very few connections comparatively. Check those longer wires. They might be off by one or two spots.

- With this project, I had different results with different speakers. If you get the one I suggest, then it should work out. If not, you might want to try a different speaker.

Thank you!

How did you enjoy this book? Please let us know. Take a moment and email us at support@pragprog.com with your feedback. Tell us your story and you could win free ebooks. Please use the subject line "Book Feedback."

Ready for your next great Pragmatic Bookshelf book? Come on over to https://pragprog.com and use the coupon code BUYANOTHER2021 to save 30% on your next ebook.

Void where prohibited, restricted, or otherwise unwelcome. Do not use ebooks near water. If rash persists, see a doctor. Doesn't apply to *The Pragmatic Programmer* ebook because it's older than the Pragmatic Bookshelf itself. Side effects may include increased knowledge and skill, increased marketability, and deep satisfaction. Increase dosage regularly.

And thank you for your continued support,

The Pragmatic Bookshelf

SAVE 30%!
Use coupon code
BUYANOTHER2021

Distributed Services with Go

This is the book for Gophers who want to learn how to build distributed systems. You know the basics of Go and are eager to put your knowledge to work. Build distributed services that are highly available, resilient, and scalable. This book is just what you need to apply Go to real-world situations. Level up your engineering skills today.

Travis Jeffery
(258 pages) ISBN: 9781680507607. $45.95
https://pragprog.com/book/tjgo

Explore Software Defined Radio

Do you want to be able to receive satellite images using nothing but your computer, an old TV antenna, and a $20 USB stick? Now you can. At last, the technology exists to turn your computer into a super radio receiver, capable of tuning in to FM, shortwave, amateur "ham," and even satellite frequencies, around the world and above it. Listen to police, fire, and aircraft signals, both in the clear and encoded. And with the book's advanced antenna design, there's no limit to the signals you can receive.

Wolfram Donat
(78 pages) ISBN: 9781680507591. $19.95
https://pragprog.com/book/wdradio

Genetic Algorithms in Elixir

From finance to artificial intelligence, genetic algorithms are a powerful tool with a wide array of applications. But you don't need an exotic new language or framework to get started; you can learn about genetic algorithms in a language you're already familiar with. Join us for an in-depth look at the algorithms, techniques, and methods that go into writing a genetic algorithm. From introductory problems to real-world applications, you'll learn the underlying principles of problem solving using genetic algorithms.

Sean Moriarity

(242 pages) ISBN: 9781680507942. $39.95

https://pragprog.com/book/smgaelixir

Design and Build Great Web APIs

APIs are transforming the business world at an increasing pace. Gain the essential skills needed to quickly design, build, and deploy quality web APIs that are robust, reliable, and resilient. Go from initial design through prototyping and implementation to deployment of mission-critical APIs for your organization. Test, secure, and deploy your API with confidence and avoid the "release into production" panic. Tackle just about any API challenge with more than a dozen open-source utilities and common programming patterns you can apply right away.

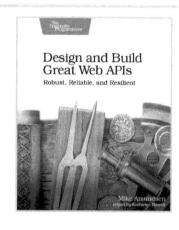

Mike Amundsen

(330 pages) ISBN: 9781680506808. $45.95

https://pragprog.com/book/maapis

Quantum Computing

You've heard that quantum computing is going to change the world. Now you can check it out for yourself. Learn how quantum computing works, and write programs that run on the IBM Q quantum computer, one of the world's first functioning quantum computers. Develop your intuition to apply quantum concepts for challenging computational tasks. Write programs to trigger quantum effects and speed up finding the right solution for your problem. Get your hands on the future of computing today.

Nihal Mehta, Ph.D.
(580 pages) ISBN: 9781680507201. $45.95
https://pragprog.com/book/nmquantum

A Common-Sense Guide to Data Structures and Algorithms, Second Edition

If you thought that data structures and algorithms were all just theory, you're missing out on what they can do for your code. Learn to use Big O Notation to make your code run faster by orders of magnitude. Choose from data structures such as hash tables, trees, and graphs to increase your code's efficiency exponentially. With simple language and clear diagrams, this book makes this complex topic accessible, no matter your background. This new edition features practice exercises in every chapter, and new chapters on topics such as dynamic programming and heaps and tries. Get the hands-on info you need to master data structures and algorithms for your day-to-day work.

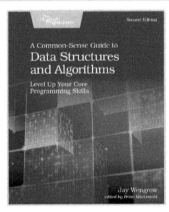

Jay Wengrow
(506 pages) ISBN: 9781680507225. $45.95
https://pragprog.com/book/jwdsal2

Build Location-Based Projects for iOS

Coding is awesome. So is being outside. With location-based iOS apps, you can combine the two for an enhanced outdoor experience. Use Swift to create your own apps that use GPS data, read sensor data from your iPhone, draw on maps, automate with geofences, and store augmented reality world maps. You'll have a great time without even noticing that you're learning. And even better, each of the projects is designed to be extended and eventually submitted to the App Store. Explore, share, and have fun.

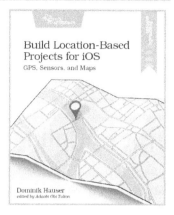

Dominik Hauser
(154 pages) ISBN: 9781680507812. $26.95
https://pragprog.com/book/dhios

iOS Unit Testing by Example

Fearlessly change the design of your iOS code with solid unit tests. Use Xcode's built-in test framework XCTest and Swift to get rapid feedback on all your code — including legacy code. Learn the tricks and techniques of testing all iOS code, especially view controllers (UIViewControllers), which are critical to iOS apps. Learn to isolate and replace dependencies in legacy code written without tests. Practice safe refactoring that makes these tests possible, and watch all your changes get verified quickly and automatically. Make even the boldest code changes with complete confidence.

Jon Reid
(358 pages) ISBN: 9781680506815. $47.95
https://pragprog.com/book/jrlegios

The Pragmatic Bookshelf

The Pragmatic Bookshelf features books written by professional developers for professional developers. The titles continue the well-known Pragmatic Programmer style and continue to garner awards and rave reviews. As development gets more and more difficult, the Pragmatic Programmers will be there with more titles and products to help you stay on top of your game.

Visit Us Online

This Book's Home Page
https://pragprog.com/book/catimers
Source code from this book, errata, and other resources. Come give us feedback, too!

Keep Up to Date
https://pragprog.com
Join our announcement mailing list (low volume) or follow us on twitter @pragprog for new titles, sales, coupons, hot tips, and more.

New and Noteworthy
https://pragprog.com/news
Check out the latest pragmatic developments, new titles and other offerings.

Save on the ebook

Save on the ebook versions of this title. Owning the paper version of this book entitles you to purchase the electronic versions at a terrific discount.

PDFs are great for carrying around on your laptop—they are hyperlinked, have color, and are fully searchable. Most titles are also available for the iPhone and iPod touch, Amazon Kindle, and other popular e-book readers.

Send a copy of your receipt to support@pragprog.com and we'll provide you with a discount coupon.

Contact Us

Online Orders:	*https://pragprog.com/catalog*
Customer Service:	*support@pragprog.com*
International Rights:	*translations@pragprog.com*
Academic Use:	*academic@pragprog.com*
Write for Us:	*http://write-for-us.pragprog.com*
Or Call:	+1 800-699-7764